YOUR KNOWLEDGE HAS VALUE

- We will publish your bachelor's and master's thesis, essays and papers

- Your own eBook and book - sold worldwide in all relevant shops

- Earn money with each sale

Upload your text at www.GRIN.com and publish for free

Bibliographic information published by the German National Library:

The German National Library lists this publication in the National Bibliography; detailed bibliographic data are available on the Internet at http://dnb.dnb.de .

This book is copyright material and must not be copied, reproduced, transferred, distributed, leased, licensed or publicly performed or used in any way except as specifically permitted in writing by the publishers, as allowed under the terms and conditions under which it was purchased or as strictly permitted by applicable copyright law. Any unauthorized distribution or use of this text may be a direct infringement of the author s and publisher s rights and those responsible may be liable in law accordingly.

Imprint:

Copyright © 2017 GRIN Verlag
Print and binding: Books on Demand GmbH, Norderstedt Germany
ISBN: 9783668647992

This book at GRIN:

https://www.grin.com/document/412079

Tugfan Sahin, S. Can

Oil Pollution Control Technologies in Turkey. EMSA-CleanSeaNet Service

GRIN Verlag

GRIN - Your knowledge has value

Since its foundation in 1998, GRIN has specialized in publishing academic texts by students, college teachers and other academics as e-book and printed book. The website www.grin.com is an ideal platform for presenting term papers, final papers, scientific essays, dissertations and specialist books.

Visit us on the internet:

http://www.grin.com/

http://www.facebook.com/grincom

http://www.twitter.com/grin_com

Pollution control technologies in Turkey with the contribution of EMSA-CleanSeaNet service

T. Sahin[1,2] and S. Can[3]

[1]Department of Maritime Transportation Engineering, Graduate School of Science Engineering and Technology, Istanbul Technical University, 34940, Istanbul, Turkey
[2]Turkish Maritime Education Foundation, 34940, Istanbul, Turkey
[3]Istanbul Technical University, Maritime Faculty, 34940, Istanbul, Turkey

Abstract 2
1. INTRODUCTION 2
   1.1.Technical specifications of the CleanSeaNet service 2
   1.2 Data collection for oil spill detections 3
2. OIL SPILL DETECTION TECHNOLOGIES IN TURKEY.. 3
3. CONCLUSIONS 10
References 11

Abstract

The importance of identifying and tracing oil pollution on the sea surface is an important fact in order to protect our environment. CleanSeaNet which is one of the European Maritime Safety Agency's services is a significant tool and service for the States to increase the environmental safety and prevent marine pollution. By using the CleanSeaNet service which is supported by satellite technology, the detection and tracing of marine pollution is quite easier and more efficient comparing to previous years. In this study, the CleanSeaNet and its practices in Turkey regarding to identifying and tracing oil pollution have been examined. Furthermore, the practices in Turkey are revealed with a few real-case examples which have been detected by means of pollution control technology; the CleanSeaNet. This paper concludes with the recommendations on future directions on how to identify and trace oil pollution by using new technologies.

Keywords: Oil spill; marine pollution; pollution control technologies; EMSA; CleanSeaNet.

1. INTRODUCTION

The CleanSeaNet is one of the (European Maritime Safety Agency) EMSA's significant tool which is a European satellite-based oil spill and vessel detection service. It offers assistance to participating States for identifying and tracing oil pollution on the sea surface, monitoring accidental pollution during emergencies and contributing to the identification of polluters. The service is based on the radar images obtained from Synthetic Aperture Radar (SAR) satellites [1].

1.1. Technical specifications of the CleanSeaNet service

There are several satellites which have been used for this purpose such as ENVISAT, RADARSAT 1 and 2, SENTINEL-1, TerraSAR-X. Besides, there are service providers which are eGEOS, CLS, Edisoft, KSAT, and MDA [2]. These service providers have the ability of reception and analysis of high resolution satellite radar images. It is a reference for maritime surveillance and it does provides satellite services in near real time. Additionally, the CleanSeaNet has the capacity to acquire image segments from 200 km long up to 1400 km with a nominal "Near Real Time" performance of 30 minutes for a 400 km long acquisition. Next, the product processing and alert generation have been carried out by EMSA CleanSeaNet Data Centre. Finally, the detection results are reported to the affected coastal state approximately 30 minutes after the satellite image acquisition although the exact time varies according to the size of the image. A total of 27 countries (23 EU Coastal States, Iceland, Norway, Montenegro, and Turkey) have been using the CleanSeaNet service [3]. The service is free of charge providing that the Condition of Use has to be signed by the member states. By signing the agreement, member states are obliged to provide information (feedback) regarding the verification of possible oil spills reported by the CleanSeaNet, to ensure follow-up and provide information on spills that were not reported.

1.2 Data collection for oil spill detections

In accordance with the EMSA (European Maritime Safety Agency), the operations within the first three years have demonstrated that the CleanSeaNet is quite efficient for the detection of oil spills. Between 16th April 2007 and 31st December 2009, a total of 5816 images have been successfully delivered to 26 coastal states which already use the CleanSeaNet service. Among these images a total of 7193 possible spills have been detected by the CleanSeaNet. 1997 have been verified on site by the member states and 542 have been confirmed as being mineral oil. The overall rate of confirmation is better than 50% if the spill is checked by aircraft no later than 3 hours after the satellite acquisition [4]. As it is shown in this Figure 1 below, the number of oil spill detections in 2014 is lower than the detections in 2008.

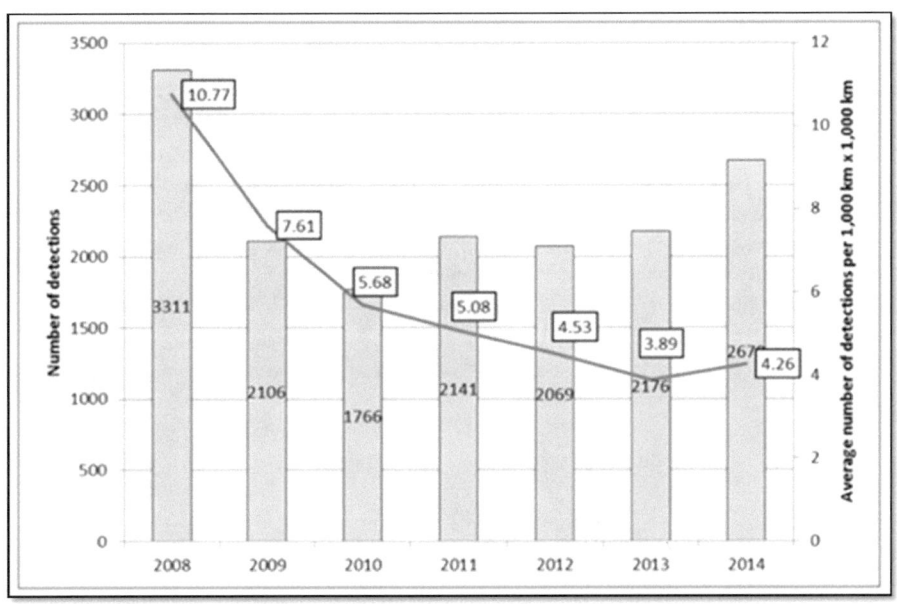

Figure 1. Number of oil spill detections between 2008 and 2014 [5].

2. OIL SPILL DETECTION TECHNOLOGIES IN TURKEY

The cooperation with the EMSA has been first initiated by the MONINFO project funded by the Black Sea Commission, which is operated within the Bucharest Convention (Protection of the Black Sea against Pollution). MONINFO is a 2 years (2009 – 2010) project, and is approved by the European Parliament (EP) and funded by the European Commission (EC). By the MONINFO Project, the efforts have been made to develop cooperative means between the riparian countries to reduce the risks of ship-based oil pollution in the Black Sea. One of the most important facts of these instruments is the cooperation work with EMSA to monitor marine pollution by means of the satellites [6]. Besides, permanent secretariat of the MONINFO Project is located in Istanbul, Turkey [7].

The Figure 2 shows a real case example in which M/V OLGA-1 vessel has been reported by Georgia, and the vessel has been inspected in Turkey resulting with a detention.

CSN Image	Notification	Ship Name Flag	Inspection Date	Inspection Place	Results
22/09/2010 19:06:39 UTC	Georgia	OLGA-1	30.09.2010	KROMAN PORT	Illegal pipeline discovered from bilge water tank to overboard, number of tanks don't match IOPP Cert., the vessel was detained on 30/09 and released on 02/10

Figure 2. Real case example within the MONINFO Project [8].

In Turkey, the CleanSeaNet service has been launched in 2011. Oil spill detections are under the responsibility and authority of the Ministry of Transport, Maritime Affairs and Communications. Additionally, the delegation of authority has been given to the Coast Guard, Istanbul Metropolitan Municipality, Kocaeli Metropolitan Municipality, Antalya Metropolitan Municipality and Mersin Metropolitan Municipality. The area of jurisdiction for detection and pollution prevention by Istanbul Metropolitan Municipality is illustrated in Figure 3.

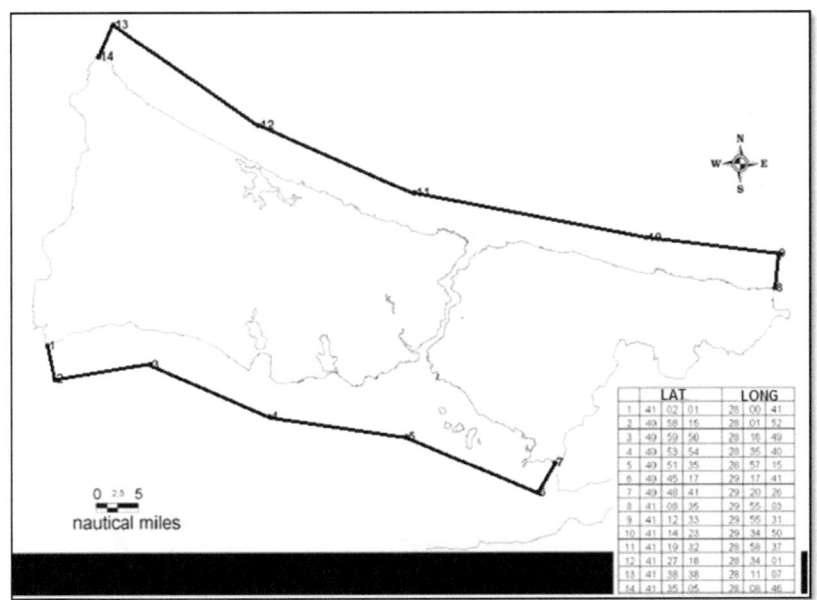

Figure 3. The coordinates and area of jurisdiction for detection oil spill and pollution prevention by Istanbul Metropolitan Municipality [9].

It is possible to report any suspicious threat within the country's territorial waters to the Coast Guard via 158, to the Provincial Directorate of Environment and Urbanisation within the region (Ministry of Environment and Urbanisation) via 181, and to the Harbour Master within the region by individuals. In accordance with the CleanSeaNet service, all reports have been collected by AAKKM (Search and Rescue and Coordination Centre) within the scope of the Ministry of Transport, Maritime Affairs and Communications. The AAKKM tracks the suspect vessel and informs the head of department for the Marine Environment and Tourism. All of the information about suspect vessels can be detected by means of AIS (Automatic Identification System) system and the information such as the name of the ship, IMO number, course, speed, destination port, ETA (Estimated Time of Arrival) to next port are identified.

Furthermore, in accordance with the instructions given by the Ministry of Transport, Maritime Affairs and Communications, oil spill which is caused by the suspect vessel can be observed on the spot by the Coast Guard boats, helicopters, unmanned air vehicles, and drones. The samplings can be taken if necessary. If the vessel's next port of call is one of the Turkish ports, the Port State Control Officers will embark and carry out inspections as soon as the vessel is berthed. Such vessels are subject to detention and administrative fines in accordance with the local law.

There is a real case about detection oil spill by using the CleanSeaNet service. In 2010, by means of the satellites of CleanSeaNet service, the Liberian flagged vessel M/V Humboldt Bay has been detected by Italian Coast Guard while she is causing to pollution in the south of Messina Strait. The satellite image of oil spill is illustrated in Figure 4.

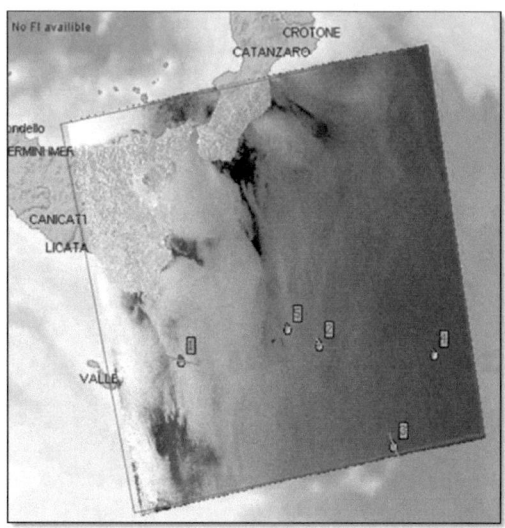

Figure 4. Satellite image of oil spill caused by M/V Humboldt Bay [10].

Since the vessel's destination port was Mersin, Turkey, the Turkish Authorities were informed and Turkish Port State Control Officers embarked the vessel for inspection in the port of Mersin. During the inspections, the following deficiencies were found onboard M/V Humboldt Bay; the entries in the oil record book was found not true, there was no IOPP (International Oil Pollution Prevention Certificate) certificate onboard, magic pipe (illegal connections) was detected between sludge and water tanks, oily water separator was found that was directly discharging overboard. Next, the Italian authorities were also informed about the process that was applied for this vessel [10].

Within the scope of the CleanSeaNet service, possible oil spills have been detected in Turkey by using RADARSAT-2 satellite and the alert reports are illustrated in Figure 5 and Figure 6.

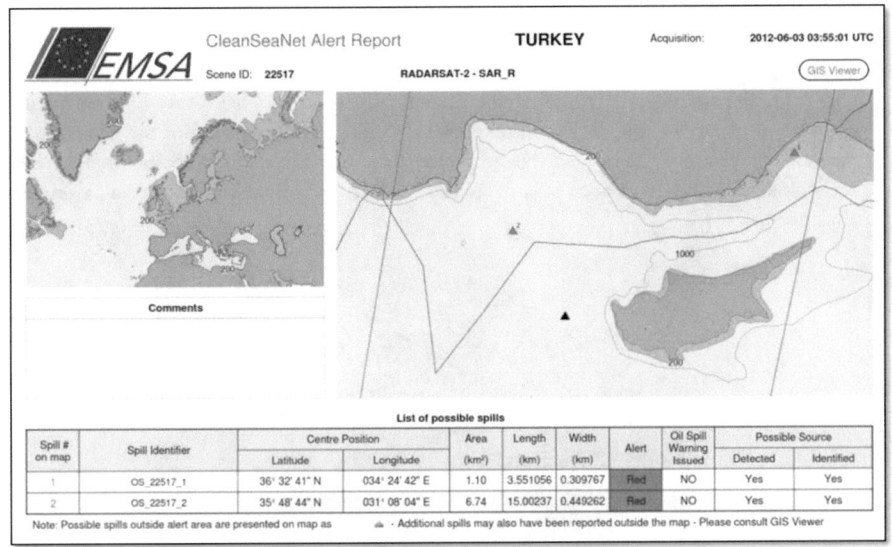

Figure 5. CleanSeaNet Alert Report [12].

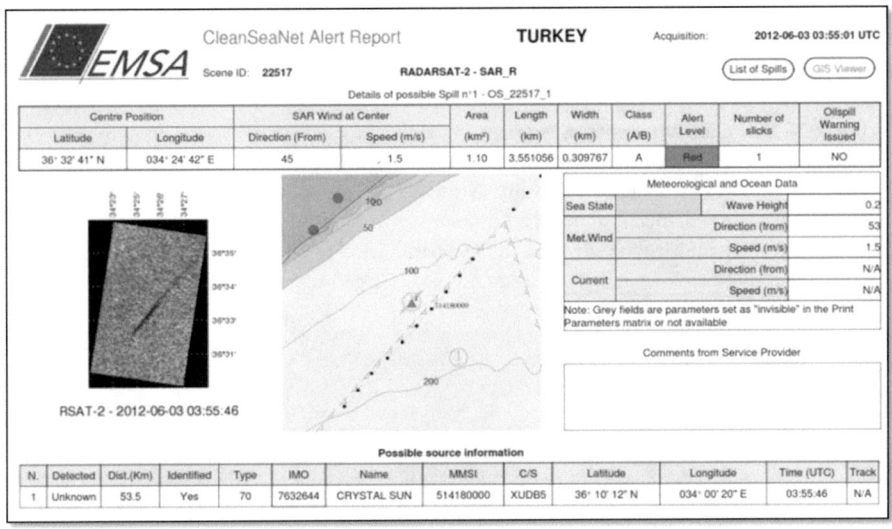

Figure 6. CleanSeaNet Alert Report [12].

According to the Ministry of Transport, Maritime Affairs and Communications, the number of vessels passing through the Istanbul Strait in 2016 is 42553. There are extra oil spill detection and control technologies in Istanbul which has a vital importance related to the number of vessels engaged in. The Metropolitan Municipality of Istanbul has been significantly tracing the ships in the vicinity and fines the vessels if there is any pollution caused by ships. In 2016, 47 vessels have been fined with a total amount of 1.200.000 Turkish Lira [11]. In particular, the Metropolitan Municipality of Istanbul has a very high technology in order to detect oil spill in its territory. Not only the satellite system of the CleanSeaNet service but also four fast-speed boats, helicopters, unmanned air vehicles such as high technological drones have been effectively used. In Figure 7, oil spill photo taken by drone is illustrated.

Figure 7. *Oil Spill Detection by Drone.* Digital image. Metropolitan Municipality of Istanbul. Web. 17 January 2017.

A 7-24 hour continuous control has been maintained including official holidays by using two separate control teams at the same time. Surveillance drones have been started to deploy by the Istanbul Metropolitan Municipality to prevent further pollution of the Istanbul Strait from the ships. The department of Marine Environment of Istanbul Municipality has also started to access the CleanSeaNet system via the Ministry. The Figure 8 and Figure 9 illustrate more examples of oil spill detections by using EMSA's CleanSeaNet programme-GIS Viewer.

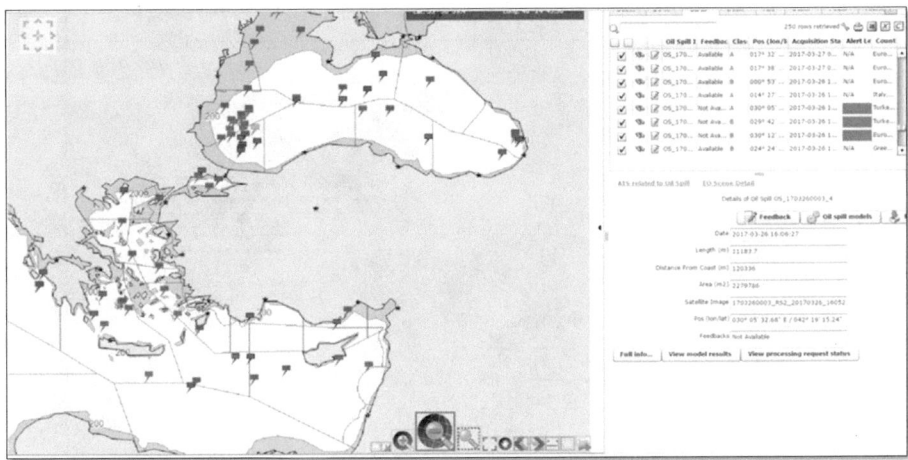

Figure 8. Oil spills detected by the CleanSeaNet in Turkey region [12].

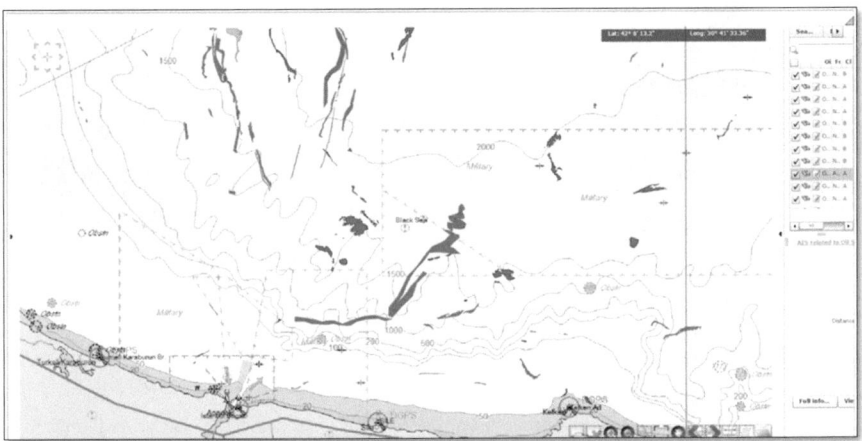

Figure 9. Oil spills detected by the CleanSeaNet in north of Istanbul Strait [12].

Last but not least, the procedures and steps which Istanbul Metropolitan Municipality follows for oil spill are shown in Figure 10.

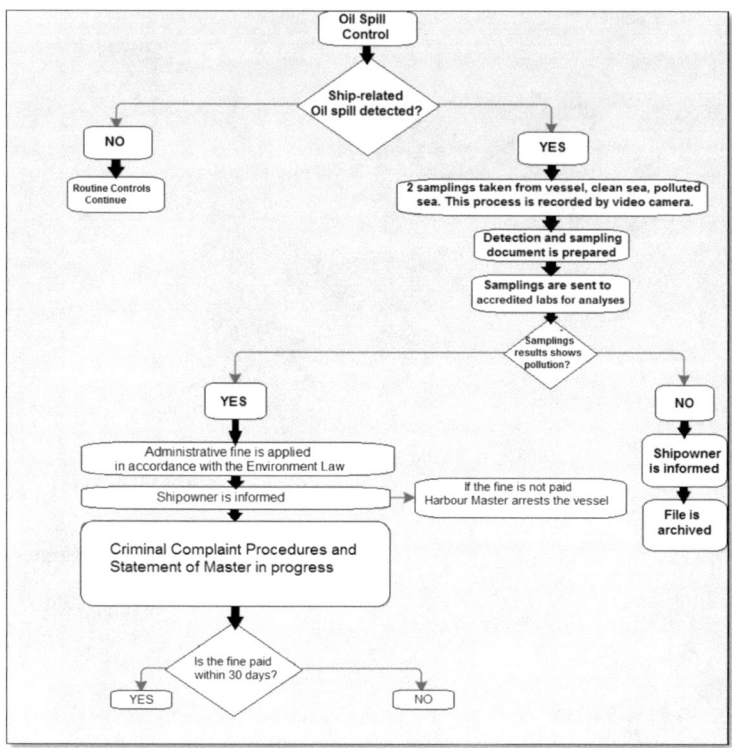

Figure 10. The procedures of Istanbul Metropolitan Municipality for oil spill event [11].

3. CONCLUSIONS

While technology improves, the prevention methods for pollution have also been diversifying. The most efficient ways to detect oil spill caused by ship wreckage and prevent pollution should be selected carefully in such a way that authorities can reach their aim effectively. It can be concluded with the recommendations on future directions in how to identify and trace oil slick by using new technologies. The number of unmanned air vehicles (drones) in operation should be increased and these equipments should be frequently used in combination with higher specific unmanned air vehicles which can operate in severe meteorological conditions as explained in this paper. When there is no data received from the CleanSeaNet system, a continuous control can be maintained by using unmanned air vehicles.

References

1. http://www.emsa.europa.eu/csn-menu.html (accessed April 4, 2017).
2. G. Hajduch, 2016. Cedre Information Day. *Vigisat ground receiving station and EMSA CleanSeaNet services,* Oct 13.https://wwz.cedre.fr/content/download/8686/138159/file/JI-2016-07-Guillaume-Hajduch-Cls.pdf (accessed April 4, 2017).
3. EMSA, 2014. *Pollution Preparedness and Response-Report 2013.*
4. http://www.emsa.europa.eu/emsa-homepage/122-operational-tasks/earthobservationservices/482-the-cleanseanet-service.html (accessed April 5, 2017).
5. M. Journel, CleanSeaNet products and requirements. *Presentation* https://www.bonnagreement.org/site/assets/files/16918/emsa_marc_journel_cleanseanet.pdf (accessed April 5, 2017).
6. The Ministry of Transport, Maritime Affairs and Communications http://www.ubak.gov.tr/BLSM_WIYS/DENIZCILIK/tr/Belgelik/20140405_145507_8109 7_1_81100.html (accessed April 10, 2017).
7. http://www.blacksea-commission.org/_convention.asp (accessed April 12, 2017).
8. Dorogan D., 2011. Black sea commission activities on environmental safety aspects of shipping MONINFO project. *International Conference, The EU Strategy for Black Sea and the Energy Policy in the Black Sea Region*
9. Figure 3. *The coordinates and area of jurisdiction for detection oil spill and pollution prevention by Istanbul Metropolitan Municipality* https://denizhizmetleri.ibb.gov.tr/neler-yapiyoruz/deniz-kirliligi-denetimi/ (accessed April 12, 2017).
10. EMSA Workshop, 2011. Agenda item 5, National Examples, Italian Coast Guard.
11. Istanbul Metropolitan Municipality website https://denizhizmetleri.ibb.gov.tr/insansiz-hava-araclari-iha-ile-deniz-denetimi/ (accessed April 5, 2017).
12. The CleanSeaNet System, *EMSA GIS Viewer* https://portal.emsa.europa.eu/home (accessed April 14, 2017).

YOUR KNOWLEDGE HAS VALUE

- We will publish your bachelor's and master's thesis, essays and papers

- Your own eBook and book - sold worldwide in all relevant shops

- Earn money with each sale

Upload your text at www.GRIN.com and publish for free